Mohammed Nadhir MEBARKI

Agricultural soil compaction

Mohammed Nadhir MEBARKI

Agricultural soil compaction

Mechanization and soil compaction

ScienciaScripts

Imprint
Any brand names and product names mentioned in this book are subject to trademark, brand or patent protection and are trademarks or registered trademarks of their respective holders. The use of brand names, product names, common names, trade names, product descriptions etc. even without a particular marking in this work is in no way to be construed to mean that such names may be regarded as unrestricted in respect of trademark and brand protection legislation and could thus be used by anyone.

Cover image: www.ingimage.com

This book is a translation from the original published under ISBN 978-613-8-49347-1.

Publisher:
Sciencia Scripts
is a trademark of
Dodo Books Indian Ocean Ltd. and OmniScriptum S.R.L publishing group

120 High Road, East Finchley, London, N2 9ED, United Kingdom
Str. Armeneasca 28/1, office 1, Chisinau MD-2012, Republic of Moldova, Europe
Printed at: see last page
ISBN: 978-620-8-26654-7

TABLE OF CONTENTS

INTRODUCTION

For half a century, Algeria has been experiencing a strong demographic dynamic which has upset various economic, social and natural balances, and has put great pressure on available agricultural resources. As a result, the country is currently experiencing a major imbalance in its agricultural trade balance, leading to extremely high levels of food dependency on foreign countries, particularly for cereals. **(Chabane and Boussard, 2012).**

Over the last 50 years, farming practices have continued to intensify with the aim of increasing productivity and minimising production costs. This includes the massive use of inputs (fertilisers and pesticides) and increased mechanisation (more powerful machinery, more intensive tillage). As a result, the soil is subject to a number of threats that are thought to lead to a deterioration in its agronomic and environmental functions. These include a reduction in organic matter levels and biodiversity, and an increase in erosion and compaction of agricultural soils **(Lefebvre, 2010).**

With the exception of overgrazed meadows, soil compaction is due to the mechanisation of farming and forestry, which involves heavy machinery passing through wet conditions. Along with erosion, this phenomenon is a major process in the physical degradation of soils in Europe. An area of 33 million hectares is currently affected, or 4% of the land area **(Amara et al., 2008)**, compared with 12% for water erosion.

The impact of soil compaction varies greatly depending on the type of cropping system. In arable farming, the pressures exerted are low (30 kPa) during sowing or soil preparation. They are very high during harvesting and transportation (150 or even 200 kPa). In vineyards, the pressures are average. (60 kPa) to high (100 kPa) pressures, but some large aircraft can have very high pressures (280 kPa). In forests, the machines used exert medium (60 kPa) to very high (190 kPa) pressures when felling or skidding timber **(Roger- Estrade et al., 2011)**.

In recent years, there has been renewed interest in this problem. Indeed, the deterioration in the quality of agricultural soils is a risk that now affects all soils in the world (due to the widespread use of mechanisation) and which, in addition to reducing production, leads to an increase in the environmental impact of agriculture: increased nitrate leaching, nitrous oxide emissions that contribute to climate change, runoff and soil erosion **(Bessou et al., 2010).**

In addition, current developments in cropping systems are putting compaction back at the heart of concerns. In arable farming, for example, there is a trend towards less tillage, motivated by a desire to reduce production costs and mechanisation costs, to protect the soil from erosion or to encourage carbon storage by reducing the rate at which organic matter is mineralised. However, this can only be done without taking too many risks by controlling soil compaction. In fact, by simplifying tillage, we limit the possibilities of regenerating the structure through fragmentation by tools. Finally, there are concerns about the irreversibility of soil degradation caused by repeated tillage in wet conditions **(Guy, 2012).**

Two types of agricultural soil compaction can be distinguished: compaction induced by the passage of agricultural machinery (compaction of the soil under the passage of wheels) and compaction induced by tillage techniques (ploughing). In both cases, the main factors influencing compaction are: soil properties (moisture content, texture and structure), soil type and soil type, and organic matter content), machine characteristics (weight, size and number of tyres), and machine operating conditions such as driving speed and tyre pressure **(Nawaz et al., 2013).** Soil compaction is currently favoured by agricultural mechanisation, with the use of increasingly heavy farm machinery and the introduction of cropping systems requiring harvesting and spreading during the wet season **(Batey 2009; Schjønning et al., 2018).**

Heavy agricultural machinery damages soil aggregates, increases soil density and moisture content, and decreases soil porosity and permeability **(Aikins and Afuakwa, 2012; Somasundaram et al., 2018)**. The main factors contributing to soil compaction are heavy wheel loads and the number of tractor passes over the field **(Augustin et al., 2020)**. Soil compaction compromises the aeration and water-holding capacity of the soil, and induces changes in its chemical and biological properties **(Ernst and Emmerling, 2013; Moinfar et al., 2021)**, leading to soil degradation and reduced crop yields **(Wozniak, 2020**
Derpsch and Friedrich, 2009; Shah et al., 2017). The harmful effects of soil compaction are not only visible in the topsoil. Deeper soil layers are also compacted, which can lead to the formation of plough footings that are very difficult to remove **(Botta et al., 2018)**. In contemporary agriculture, efforts are being made to replace energy-intensive ploughing with simplified ploughing, to reduce the number and intensity of ploughing and loosening operations, or even to eliminate these practices altogether **(Orzech and Załuski, 2020).** However, the adoption of no-till systems in the long term, can also cause compacted layers

up to 20 cm, characterised by high bulk density, low porosity and high mechanical strength. In this context, and in order to promote the development of sustainable agriculture, the combined effects of soil compaction and conventional and simplified tillage should be taken into account. of the soil, should be studied with the utmost care, in order to design cropping systems that maximise yields and minimise soil degradation. This book discusses the evaluation of the effects of soil compaction and different tillage systems on bulk density, soil water content, penetrometric resistance, infiltration rate and microbial activity in the soil, and the consequences for the development of barley and durum wheat sown in a Mediterranean climate in different soil textures.

This work is organised around two chapters: **Chapter 01**, which deals with the study of compaction of agricultural soils in the laboratory. The aim is to analyse the effect of compaction intensity on soil structure and its consequences on the root development of three durum wheat varieties, as well as a study of soil compressibility as a function of moisture variation.

Chapter 02 of the experimentation part focused on the analysis of the combined effect of soil compaction and cultivation techniques (conventional tillage and direct seeding) on the physico-mechanical behaviour of the soil and its impact on the morphological characteristics of barley, as well as on root development and yield. This experimentation was carried out at the ITGC in Oued Smar, Algiers.

CHAPTER 1
LABORATORY STUDY OF THE COMPACTION OF AGRICULTURAL SOILS

Introduction

The reduced productivity of cereals on compacted soil is mainly the result of reduced root elongation rates and therefore limited access to water and nutrients, where in compacted soils root growth is reduced due to high resistance to soil penetration. To study the effect of compaction on the development of durum wheat and its impact on the root system, we have carried out two laboratory tests in this chapter. The choice of this crop is dictated by the fact that wheat and cereal production in Algeria is not overly demanding.

1.1.Materials used

Specific equipment was used to carry out this experiment. The trials were carried out inside an agricultural greenhouse.

1.1.1. Plant material

The crop chosen for this experiment was durum wheat (three varieties), which was chosen because of its well-developed root system. The first variety of durum wheat (Triticumdurum, Desf) to be used is the mohammed ben bachir (MBB) variety. It has been a local variety since 1931, originating from the Central Station for the Improvement of Field Crops (ITGC 1994). It is the result of a genealogical selection carried out on a variety from the northern Sétif region **(Kadra, 1974)**. It has high, hollow straws and light amber grains of medium size. It has a late development cycle, average tillering and average productivity. The second durum wheat variety used is Bousselam, a Heider/MT/Ho cross from the International Center for Agricultural Research in the Dry Areas (**ICARDA**). It is a selection from the ITGC seed farm (Sétif). It is productive and measures 90-100 cm in height, with a high GMP. Its ears are semi-loose, white with black barbs and moderately sensitive to shattering. The third variety of durum wheat used, Boutaleb (Hedba 3 X ofanto), originates from the ITGC central arable crop improvement station and is grown in the high plateaux and interior plain.

1.1.2. The experimental unit

We grew the wheat in pots, which we made ourselves to make it easier to observe the development of the roots. They are cubic in shape, with wooden supports and Plexiglas on the front and back. We chose Poly methyl methacrylate to enable us to appreciate the evolution of the roots. The figure below shows the pots used for our plants:

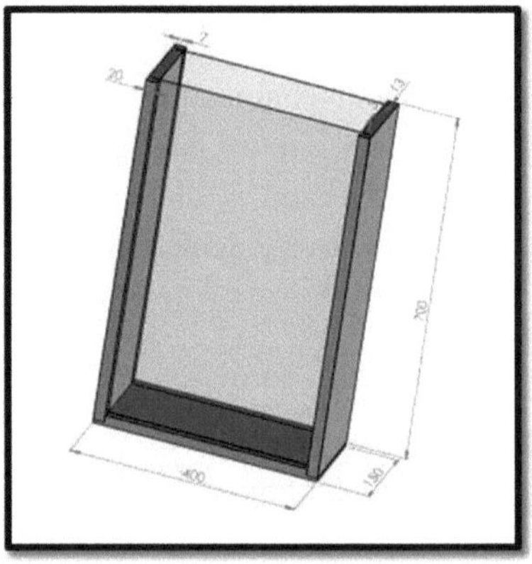

Figure 1: Plexiglas pot made using Solidworks 2016 software

Figure 2: Plexiglas pots

1.1.3. The laboratory compaction tool

We used a hydraulic press to compact the soil. The following diagram shows the tool used:

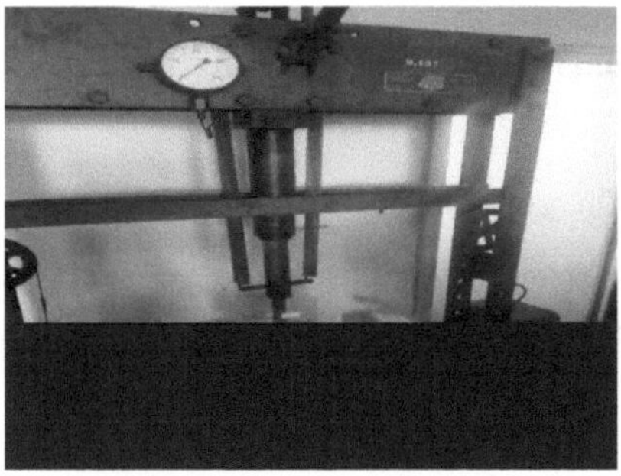

Figure 3: The hydraulic press

1.2. Methodology

Naturally, a plan was followed for this experiment, from the first stage, which involved preparing the soil samples, to the final stage, which consisted of the following the last stage where we quantified the degree of compaction and the indices of wheat and root development.

1.2.1. First step: preparing soil samples

Before sowing the wheat, we took care to prepare soil samples to fill the pots. Before preparing these samples, we chose a clay texture for the soil, and the soil analysis gave the following results:

Table 1: Analysis of the soil samples used for the test

The Parameter	Particle size					Organic matter	Total limestone	pH
	Fine silt	Silt	Sand end	Coarse sand	Clay			
The value	41.69%	22.89 %	7.67%	12.46 %	15.29 %	1.19%	1.72%	7.78
La Method	Robinson's pipette					Method	Method	pH

Our soil belongs to the **clay** textural class, with a clay content of around **15.29%** and a pH of 7.78. According to the clay content, this has a considerable influence on the soil's susceptibility to compaction. The clay content of our soil is between 10% and 22%. **(Kouassi 1977).**

1.2.2. Second stage: determining the germination rate of the three durum wheat varieties

1.2.2.1. Determining the germination rate

Wheat, like all cereals, needs good germination to grow successfully. To ensure a good crop, it is vital to know the germination capacity of the seeds sown. Determining the rate of germination in the laboratory gives us the following values for ten samples of 100 seeds.

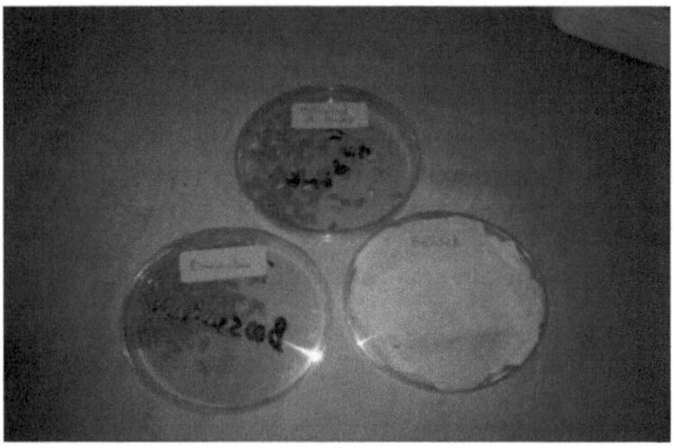

Figure 1: germination rate of the three wheat varieties

Table 2: Germination rate of 10 samples of 100 grains variety mohamed ben bachir

Samples	1	2	3	4	5	6	7	8	9	10
Rate of germination	90	87	90	91	88	90	88	86	90	91

The average germination rate is therefore **88.1%,** which is a good figure.

Table 3: germination rate of 10 samples of 100 seeds of the boutaleb variety

Samples	1	2	3	4	5	6	7	8	9	10
Rate of germination	85	84	87	90	89	88	90	89	85	84

The average germination rate is therefore **86%**, which is a good figure.

Table 4: Germination rate of 10 samples of 100 grains variety Bousselam

Samples	1	2	3	4	5	6	7	8	9	10
Rateof germination	89	88	91	90	89	90	85	89	88	86

➢ **The average germination rate is therefore 88.5%, which is a good figure.**

1.2.2. Third stage: filling the pots

This stage consisted of filling the pots while compacting the soil according to the type of compaction. We divided the pots into four groups of six pots. In the first group, we left the soil uncompacted. In the second group, we compacted the soil on the surface. In the third group, we compacted the soil at the depth of the pots, and in the fourth group, we compacted the soil completely. Each group contained six pots, which were divided into three varieties and two moisture levels.

➢ **The first group**: control pots: uncompacted soil

Figure 5: uncompacted soil

➢ **The second group**: Surface compacted soil
➢ **Third group**: deeply compacted soil
➢ **Fourth group**: fully compacted soil

2.2.3. Fourth stage: monitoring the crop

The pots were placed in a greenhouse in the Department of Rural Engineering at the Ecole Nationale Supérieure Agronomique.

Figure 6: The agricultural greenhouse in the experimental trial

1.2.4. Method for determining the length of plants and roots

It consists of measuring the elongation of the roots and stems of the plants in each pot using a calliper and a metre. Given that we were working in the greenhouse on pots and not in the field, we tried to measure the maximum rooting depth reached by the plant's roots on site by digging, which was obviously easier than in the field, especially as we had the advantage of having pots made of Plexiglas, which enabled us to better appreciate the path taken by the roots.

1.2.5. Method for determining root density

This consists of taking samples of soil with roots from a surface area of 225 cm^2 and the maximum depth of the pot. This mixture of soil and roots is placed in a fine-mesh sieve and the soil is removed using a jet of water. The samples are then left in water for 24 hours and washed again until all the soil has been removed. They are then dried and weighed.

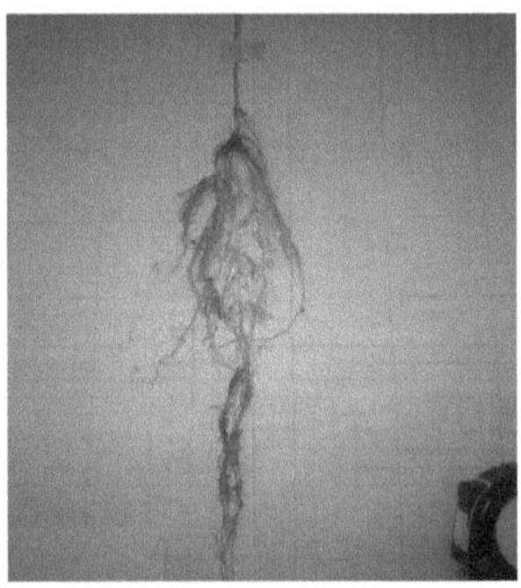

Figure 7: measuring root density

2.2.6. Choice of parameters and indices

In order to analyse the impact of compaction on plant development and the root system, we will study the plant's root development through rooting depth, root density and stem elongation. These three parameters characterise the importance of root development and will enable us to judge the quality of the roots in relation to the various degrees and types of soil compaction, at two different moisture levels and for three durum wheat varieties.

1.3. Results and interpretation

1.3.1. Analysis of the combined effect of compaction and soil moisture on the growth of three wheat varieties

Variation in stem and root height of wheat plants as a function of soil compaction and moisture.

1.3.1.1. Variation in wheat stem height as a function of compaction type and humidity (MBB variety)

The figure below shows the variation in the height of wheat stems as a function of the type of compaction and humidity (mohamed ben bachir variety):

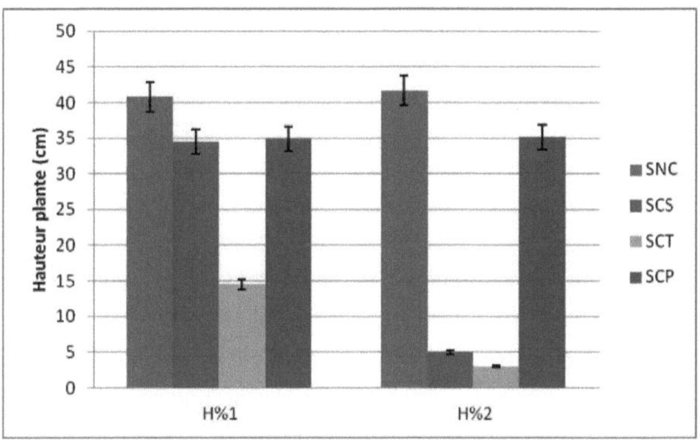

Figure 8: Variation in wheat stalk height as a function of compaction type and humidity (mohamed ben bachir variety)

Plant height development appears to differ for the three types of compaction and variation in soil moisture content. The best height of the wheat plants (mohamed ben bachir) was obtained for the first group of pots, where **no compaction** was applied. In contrast, the small heights were obtained for the second and third groups, where we applied **surface** and **total compaction** respectively. This compaction certainly had a negative effect on the development of the plant with the absence of soil refinement and the good fine soil in which the seed was placed after compaction. However, there was a big difference in the heights obtained in the second and third group pots compacted in H% 1 (dry soil), where we recorded the best results. This can only be explained by the fact that compaction has a greater effect on wet soils than dry soils.

1.3.1.2. Variation in wheat root length as a function of compaction type and humidity (mohamed ben bachir variety)

The figure below shows the variation in the length of wheat roots as a function of the type of compaction and humidity (mohamed ben bachir variety):

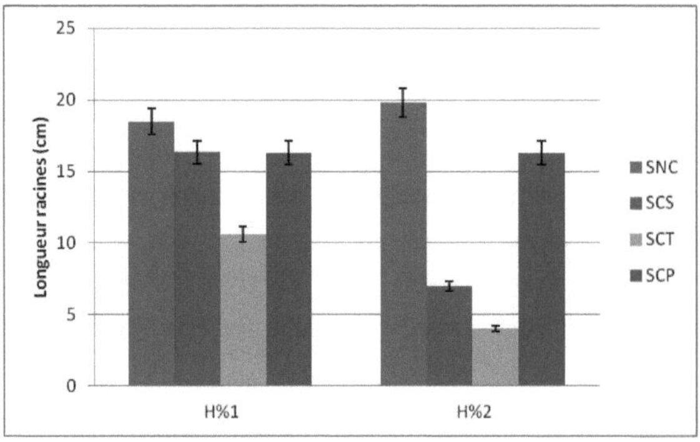

Figure 9: Variation in wheat root length as a function of compaction type and humidity (mohamed ben bachir variety)

It is clear from the results that the development of the root system of the mohamed ben bachir wheat variety seems to be different for the three cases of soil compaction and moisture content. The best root development was obtained in soils where **no compaction** was applied. Unlike the soils where we applied **surface compaction** and **total compaction**, root length. Compaction certainly had a negative effect on root development. However, there was a big difference between the root length values in the two series of second and third group pots compacted at two different humidity levels. We recorded the best results in the pots compacted in a dry state.

1.3.1.3. Variation in wheat root density as a function of compaction type and humidity (mohamed ben bachir variety)

The figure below shows the variation in root density of wheat as a function of the type of compaction and humidity (mohamed ben bachir variety):

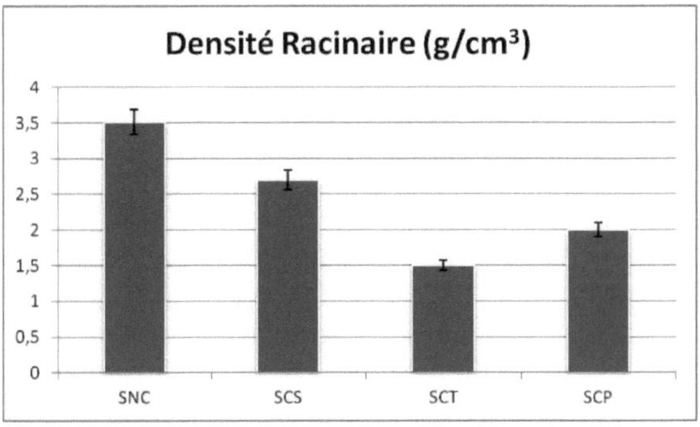

Figure 10: Variation in wheat root density as a function of compaction type and humidity (mohamed ben bachir variety)

The development of root density seems to be different for the different cases of compaction. The best wheat root densities were obtained for the first group of pots, where **no compaction** was applied**,** with a value of **3.5 g/dm^3** . For the second group of pots compacted on the surface only, we recorded an average value of **2.6 g/dm^3** . The lowest root density was obtained for the third group, where we applied **total compaction**, with a value of 1.5 g/dm$^{3.}$ We can see that there is an inverse relationship between the type of compaction and the root density. However, as the degree of compaction increases, we record small average values for root densities.

1.3.1.4. Variation in wheat stem height as a function of compaction type and humidity (boutaleb variety)

The figure below shows the variation in the height of wheat stems as a function of the type of compaction and humidity (**boutaleb variety):**

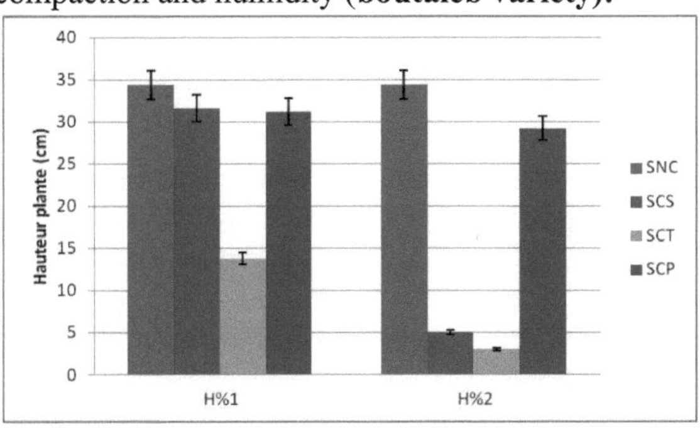

Figure 2: Variation in wheat stem height as a function of compaction type and humidity (boutaleb variety)

For the Boutaleb variety, the results obtained are similar to those for the mohamed ben bachir variety, and show that the development of plant height seems to be different for the different cases of compaction and soil humidity. The tallest wheat plants were obtained in the first group of pots, where **no compaction** was applied. On the other hand, the plants that found it difficult to gain height were obtained in the second and third groups, where we applied **surface compaction** and **total compaction** respectively. There was also a big difference in the heights obtained in the second and third group pots for the two humidity levels.

1.3.1.5. variation in wheat root length as a function of compaction type and humidity (boutaleb variety)

The figure below shows the variation in the length of wheat roots as a function of the type of compaction and humidity of the boutaleb variety:

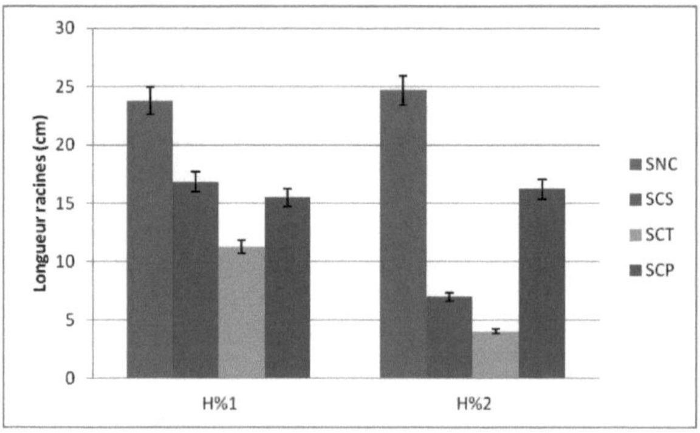

Figure 3: Variation in wheat root length as a function of compaction type and humidity (boutaleb variety)

As far as root system development is concerned, it also seems to differ for the different cases of compaction and soil moisture content. The best results were obtained for the first group of control pots without compaction, and we obtained the worst results for root development in the second and third groups, where we **applied surface compaction** and **total compaction** respectively.

1.3.1.6. The variation of wheat roots according to the type of compaction and humidity (boutaleb variety)

The figure below shows the variation in wheat roots as a function of the type of compaction and humidity of the boutaleb variety.

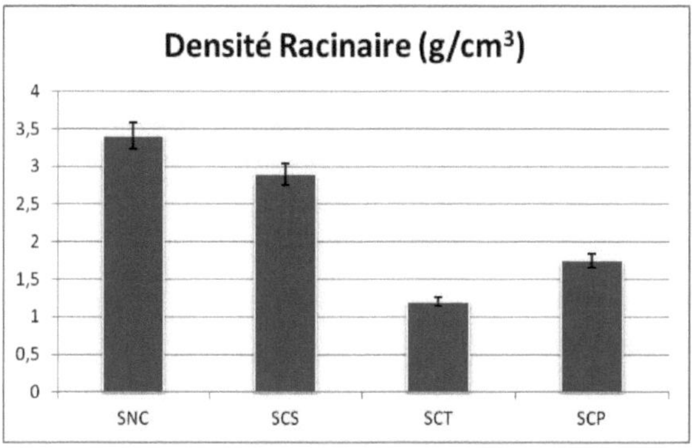

Figure 13: Variation in wheat roots as a function of compaction type and moisture (boutaleb variety)

The development of root density seems to be different for the different cases of compaction. The best root density was obtained for the first group of pots, where **no compaction** was applied**,** with a value of **3.4 g/dm^3** , we obtained the worst results for the third **totally compacted** group, with a value of **1.2 g/dm$^{3.}$**

1.3.1.7. Variation in wheat stem height as a function of compaction type and humidity (bousselam variety)

The figure below shows the variation in the height of wheat stems as a function of the type of compaction and humidity (bousselam variety):

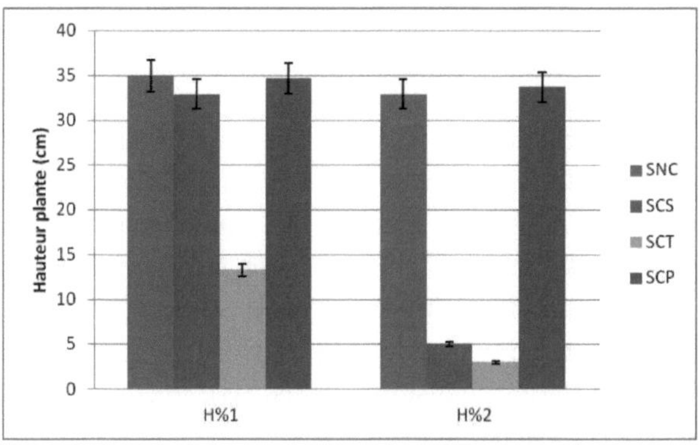

Figure 14: Variation in wheat stem height as a function of compaction type and moisture (bousselam variety)

Plant height development was difficult in compacted soils in a wet state. The best plant height for wheat (bousselam) was obtained in the first group of pots, where **no compaction** was applied. In contrast, low heights were obtained in the second and third groups, where **surface** and **total compaction were** applied respectively. However, low heights were observed in the second and third group pots compacted in H% 2 (moist soil). This can only be explained by compaction, which has a greater effect on wet soils than dry soils.

1.3.1.8. Variation in wheat root length as a function of compaction type and humidity (bousselam variety)

The figure below shows the variation in the length of wheat roots as a function of the type of compaction and humidity of the bousselam variety:

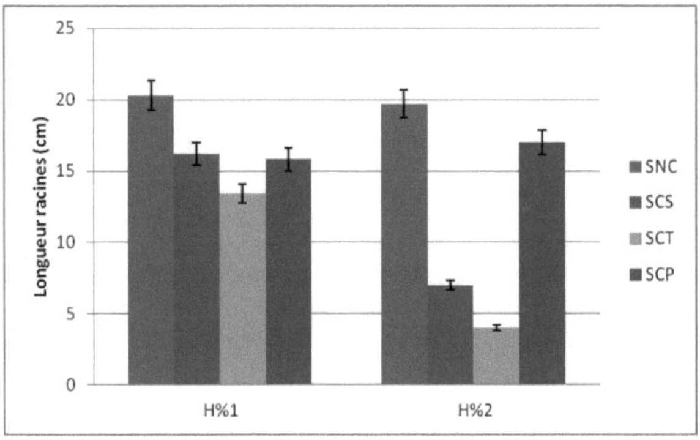

Figure 15: Variation in wheat root length as a function of compaction type and moisture (bousselam variety)

It appears from the figure above that the development of the root system of the bousselem variety seems to be best for the first group of pots, where **no compaction** was applied. The poor results were always obtained for the second and third groups, where we **compacted the moist soil on the surface** and **compacted all the soil**, respectively. This compaction certainly had a negative effect on plant root development. However, there was a large difference between the root length values in the two series of pots from the second and third groups compacted at two different moisture levels. We recorded the highest root densities in the pots compacted in the dry state.

1.3.1.9. Variation in wheat root density as a function of compaction type and humidity (bousselam variety)

The figure below shows the variation in root density of wheat as a function of the type of compaction and humidity (bousselam variety):

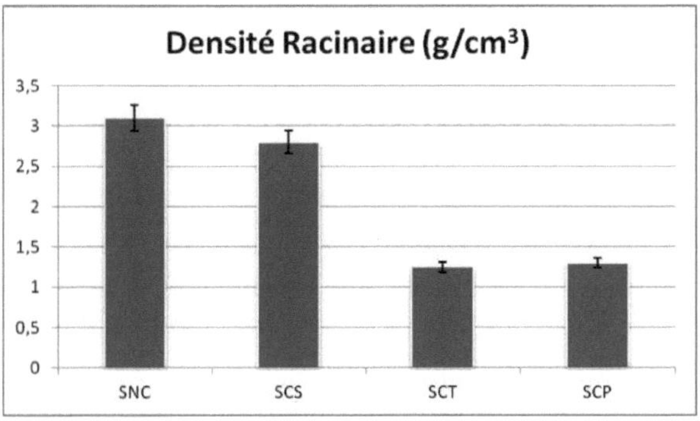

Figure 16: Variation in wheat roots as a function of compaction type and moisture (bousselam variety)

The average values of root density recorded for the different cases of compaction in dry and wet conditions give the best densities of wheat roots obtained in the first group of pots, where **no compaction** was applied**,** with a value of **3 g/dm³** . For the second group of pots compacted on the surface only, we recorded an average value of **2.6 g/dm³** . For the third group of pots compacted at depth, we recorded an average value of **2.2 g/dm³·** The lowest root density was obtained in the third group, where we **applied total compaction**. With a value of **1.1 g/dm³·** There is a proportional relationship between the type of compaction and root density, but as the degree of compaction increases, root densities decrease.

Conclusion

The results show that plant height, root depth and root density are inversely proportional to soil compaction and moisture content, i.e. the less the soil is compacted, the greater the stem height, root depth and root density. And this is what we recorded in the first group of pots (the control) with an average height of 35 cm, an average root depth of 21 cm, and a root density of **3 g/dm³** . We recorded a low average height in compacted soil (15 cm), an average root depth of 11 cm, and a root density of **1.2 g/dm³** . The results obtained also enabled us to observe the effect of compaction on wet and dry soils. We recorded satisfactory results for root heights and densities in soil compacted in a dry state compared with that compacted in a wet state. Hence the importance of choosing the right time to work with farm implements. There is also a difference between the average values recorded for plant height, root length and root density for the three wheat varieties. The **mohamed ben bachir** variety is more resistant to compaction than the other two varieties. This showed us that the choice of crop and variety is very important in the case of compacted soils. These results also enabled us to observe the effect of soil compaction and moisture on the height and root development of a plant such as durum wheat. This indirectly provides us with useful information for durum wheat sowing operations. This will obviously lead to an optimum soil structure that will allow good root development and, consequently, a good yield.

CHAPTER 2

ANALYSIS OF THE COMBINED EFFECT OF COMPACTION AND TILLAGE ON THE PHYSICO-MECHANICAL BEHAVIOUR OF A CLAY SOIL AND CONSEQUENCES FOR BARLEY DEVELOPMENT

Introduction

The aim of this trial is to study the effects of interactions between the operation of agricultural machinery (number of machines, machine weight, number of passes, etc.) and cultivation techniques (tillage and direct seeding) on soil compaction. And its consequences on the physico-chemical properties of the soil (water content, soil porosity, penetration resistance, infiltration rate and microbial activity), and on the development of the barley crop (settlement rate, root system development and yields).

2.1.Presentation of the experimental site

2.1.1. Description and location of the experimental site

The experimental trials which are the subject of our study took place at the experimental station of the Institut Technique des Grandes Cultures (ITGC) at Oued Smar in the north-eastern part of the Mitidja. The experimental station has an altitude of 24m, latitude 36°43' North, longitude 30° 84' East (**figure 17**). Generally speaking, it occupies an area of around 100 ha, with a scientific vocation, and is responsible for developing the field crop sector (cereal crops, pulses, fodder crops), in order to meet the country's food needs and the new challenges revealed by the food crises and climatic risks.

Figure 17: Geographical location of the Oued Smar technical institute for field crops (ITGC)

2.1.2. The climate of the study region

The ITGC experimental station at Oued Smar has a humid climate, with mild, rainy winters and hot, dry summers. The BAGNOULS and GAUSSEN umbrothermal diagram (Figure 29) calculated for the period from 2010 to 2019, shows the existence of a wet and relatively cold season extending over seven (07) months, and a dry and hot season for the rest of the year.

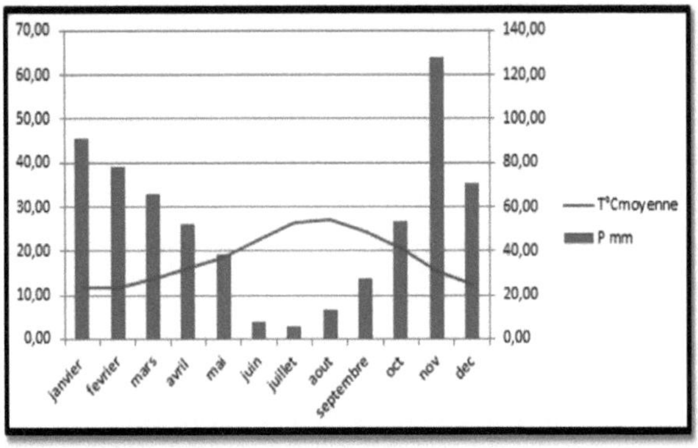

Figure 18: The BAGNOULS and GAUSSEN umbrothermal diagram (Dar El-Beida region)

With :
P: total monthly rainfall

T: average monthly temperature

2.1.3. Climatic conditions during the trial

The rainfall recorded during the trial campaign did not exceed the 100-year average for the region, at just 442.4 mm compared with 749 mm for the region as a whole, representing a 40.9% shortfall. The greatest amount of rain fell between March and May, during the bolting period. What is exceptional about this crop is that the rainfall was not well distributed throughout the plant's cycle, with a period of drought recorded in autumn and winter.

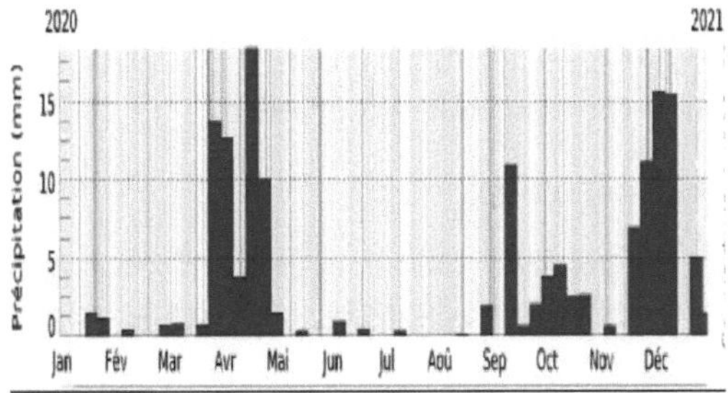

(Source: Meteoblue 2020) Figure 19: Recorded rainfall for 2020-2021

The winter period (December 2019, January and February 2020) was characterised by high temperatures, little rain and rather spring-like days, according to seasonal data from the National Meteorological Office (ONM). Seasonal cumulative rainfall was below normal, while the average seasonal temperature was above normal.

2.1.4. Soil characteristics of trial plots

Granulometric analysis of soil samples from the experimental site gave the results shown in the table below:

Table 5: Soil analysis of the test plot

The Parameter	Particle size					Organic matter	total limestone	pH
	fine silt	silt	Sand end	coarse sand	Clay			
The value	41.69%	22.89 %	7.67%	12.46 %	15.29 %	1.19%	1.72%	7.78
La Method	Robinson's pipette					Method	method	pH

Source: (INSID, 2020)

Use of the USDA texture triangle revealed that the soil in the experiment was **clay-textured.**

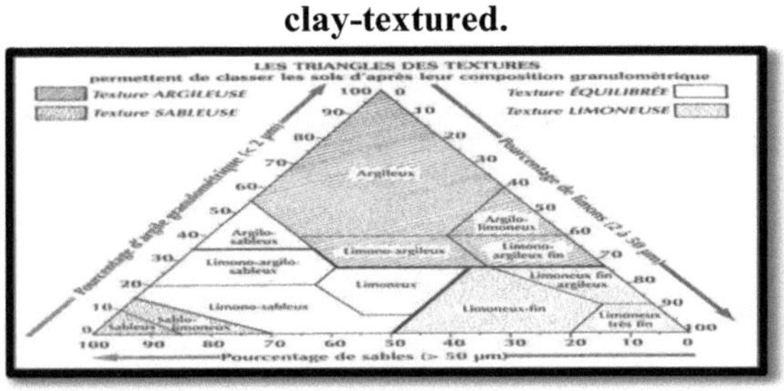

2.2. Materials and methods

2.2.1. Presentation of the experimental protocol

2.2.1.1. The experimental set-up

Taking into account the geometry and topography of the plot and the objectives of our experiment, the factorial block design was chosen as the experimental design. We have two factors :

➢ **Factor 1: Soil compaction Level 1**: Compacted soil (C) **Level 2**: Non-compacted soil (NC)
➢ **Factor 2: Cultivation technique**

Level 1: Conventional work (TC)

Level 2: Direct seeding (DS) This gives us treatments :

• **T1: compacted no-till soil**

• **T2: uncompacted direct-seeded soil**

• **T3: uncompacted conventional soil**

• **T4: uncompacted conventional soil**

With three repetitions, this gives us three blocks depending on the direction of the slope.

With :

• **T1: compacted no-till soil (SCSD)**

• **T2: uncompacted direct seeding soil (SNCSD)**

• **T3: non-compacted conventional soil (SNCTC)**

• **T4: uncompacted conventional soil (SCTC)**

2.2.2. Study materials

2.2.2.1. Soil preparation equipment

	The equipment used	Features	Illustrations
Means of traction	4*4 tractor DEUTZ-FAHR AGROLUX 80	Power: 80 hp Number of cylinders 4 Displacement : 4085 PTO speed: 540 rpm Gross Vehicle Weight Rating: 3100kg Front tyres in 4-wheel drive	
The plough Sprayer (Cover-crop)	Labour : Bisoc reversible plough	Number of ploughshares: 2 Type : range Working depth: 25 to 30 cm	
	Ploughing over	Number of discs 20 Type: trailed Diameter of the disc 55cm Weight: 740 kg	
Sowing equipment	seed drill : AGRIC PSM 30	Working width: 3m Number of elements distributors : 19 Distance between organs: 17 cm System of	
	Direct seeding drill	Working width: 3m Number of spool valves: 29. Distance between organs: 13.7	
Maintenance equipment	Roller	. Working width: 2m . Diameter: 50cm . Working speed: 6Kmh	
Plant protection equipment	Cereal sprayer with spray boom	. Capacity: 1000 L . Number of nozzles: 16 . Working width: 8m . Type of nozzles: slotted . Type : trailed . Empty weight: 302 kg	

2.2.2.2. Plant material

The plant material used in our study is barley **of the Fouara** variety. It is a 6-row barley and is the result of a selection by the Institut Technique des Grandes Cultures (ITGC) station at Sétif, using segregated material from ICARDA (International Center of Agricultural Research in Dry Areas) in 1991/92. It is a highly productive variety with good stability, and has been included in the national catalogue since 1997 **(Benmahammed et al. 2001, Menad et al. 2011).** A crop is considered to be the most important fodder crop, important for its green production, hay, silage, grain and straw.

Table 6: Technical data for Fouara barley :

Culture	Barley
Variety	Fouara
The origin	ICARDA Syria
Maximum yield	<60 qx /ha
Stem height	60-70 cm
PMG	High (36.7 g)
Protein content	Small quantities <13%.
Water content	7.4%

Source : (ITGC Oued Smar)

Figure 21: Seed of the Fouara variety

2.2.3. Conducting and monitoring the experimental study

2.2.3.1. Soil preparation and sowing

The trials were carried out in a 1245 m plot^2 which was divided into three blocks: each block divided into four micro-plots, worked and unworked, compacted and uncompacted, were worked at the end of December and sown at the beginning of January.

2.2.3.2. The size of our plot

- Total length of plot: 83 m
- Total width of parcel (length of micro-parcel): 15 m
- Total surface area of the plot: 1245 m^2.
- Width of micro-parcel: 6 m
- Surface area of micro-parcel: 90 m^2.
- Spacing between microplots: 1 m.

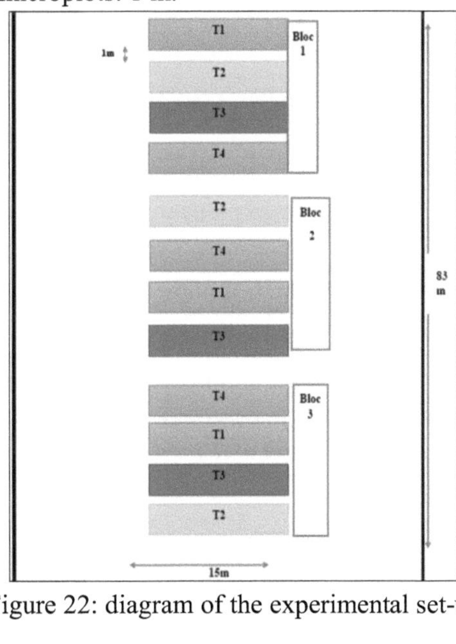

Figure 22: diagram of the experimental set-up

2.2.3.3. Conventional work

Ploughing is necessary to ensure better preparation of the seedbed, but also to improve aeration of the soil and thereby increase the space for rooting and root development.

•Ploughing using a reversible bisoc plough was carried out on 30 December 2019 in very good soil conditions, on the six microplots reserved for the technique known as conventional sowing with conventional tillage.

Table 7: Depth and type of ploughing.

Depth of ploughing (cm)	25	23	23	20	21	23	20	23	22	22	20	23	Means
observation	It's a medium-sized, flat plough												

•Ploughing plus shallow cultivation was resumed on 02 January 2020. Sowing took place on the same day for the ploughed microplots.

Figure 23: Non-compacted plot with conventional tillage

2.2.3.4. Compaction of areas on the test plot

Before sowing, we compacted the soil in the microplots concerned by the compaction, by excessive passes (five passes) of the (tractor + seed drill filled with seed), with a total axle load equal to 4 tonnes.

Figure 24: Plot after compaction

2.2.3.5. Sowing

➢ **Classic seeding**

Sowing was carried out on 02/01/2019 using an AGRIC seed drill.

Figure 25: Sowing with the seed drill

•After sowing, a roller was used in the uncompacted micro-plots to compact the soil slightly to improve contact between the soil and the seed, encourage water to rise by capillary action and prevent the surface from drying out too quickly, and to complete the action of the previous tools by reducing the size of the clods on the surface.

➢ **Application of fertiliser**

The granulated base fertiliser "M.A.P 12.52" (mono-ammonium- phosphate-granule) was applied on 02/01/2019 to the stolen crop.

Fertiliser composition: N=12%, P=52%, K = 0%.
➢ Total dose: for the total surface area: 29.5 Kg
➢ Dose per microplot: 2.5 kg

2.2.3.6. Direct Seeding

➢ **Before direct seeding: Chemical weeding**

By adopting no-till, chemical treatment is more than necessary to control weeds. The chemical treatments for the no-till plots were carried out at the end of December, on 30 December 2019, using a sprayer with a PTO speed of 540 rpm. The product used in the trial was a glyphosate-based total herbicide. Its commercial name is GLYFOZEL 36 SL, and the dose was 3L/ha, according to the manufacturer and INPV recommendations.

Figure 26: Total herbicide based on glyphosate 36 SL

2.3. Measurement techniques and methods

The analyses carried out in the laboratory and their methods are presented in the table below.

Table 8: Analyses carried out in the laboratory and their methods

Analysis		Method
	Particle size	Pipette method
	Soil moisture	Weight humidity at using an auger and
	Resistance to	Static penetrometer
Physical soil analysis	Apparent density	By coring
	Real density	Pycnometer
	Porosity	Calculation
	Conductivity	Conductivity meter
	PH	PH meter
Chemical Analysis of	Total nitrogen	KJEDHEL method

Source: (INSID, ITGC and INRA 2020)

2.3.1. Determining soil moisture

The weight moisture method was used to determine soil moisture. Samples were taken using an auger on each elementary plot of each technical itinerary. The samples were taken from four points on each plot, and each sample was taken from four horizons: from 0 to 5 cm, from 5 to 15 cm, from 15 to 25 cm and from 25 to 30 cm. The method for directly measuring the water content of a soil sample is simply to weigh it before and after steaming for 24 hours at 105 C° , these two values corresponding respectively to the total mass of the sample Pi and the mass of the solid fraction Ps.

The water content of soya is then calculated using the following formula:

$$H\% = [\frac{Pi - Ps}{Ps}]$$

With :
H%: Humidity Pi = Initial weight
Ps = Dry weight after oven drying at 105 C° for 24 hours.

Figure 27: Taking soil samples to determine soil moisture content

3.3.2. Determination of penetrometric resistance

The hand-held static penetrometer is an indicative instrument for measuring maximum penetration resistance. The instrument has an average deviation of + and - 8%. For accurate, long-term use, a range of 200 to 700 N is recommended (green zone on the gauge). Above 700N, the zone is considered to be compacted. The resistance is read in N (Newton) and noted for the appropriate depth. The base of the cone zone must also be noted, as the resistance of the cone is expressed in daN /cm^2 , or in Mpa. A static penetrometer was used to determine the depth. The method consists of applying a force to push the penetrometer rod into the ground, enabling us to measure the depth using a scale on the rod carrying the cone and read the value of the force indicated by a cursor on the penetrometer.

Figure 28: Static penetrometer

2.3.3. Determination of porosity

2.3.3.1. Determination of dry density

The most commonly used method, and the one we used, is the coring method: a cylinder of known volume, the end of which has been sharpened, is mechanically driven into the ground, then cleared by digging around it. After drying in an oven (at 105°C for 24 hours) this volume is weighed, enabling the apparent density to be calculated, which is in fact a simple ratio

$$\rho_{as} = \frac{P}{V}$$

With :

- ρas: bulk density ;
- P: weight of dry sample.
- V: volume of soil sample.

3.3.3.2. Determination of true density

To determine the real density, we used a simple method consisting of taking a sample of the soil and drying it in an oven (105°C, 24 hours). Using a graduated cylinder, we added 100g of dry soil, crushed and sieved, and added a volume of water (V0 = 800 ml), leaving the sample until all the air bubbles have disappeared. A volume of water was displaced, corresponding to the volume occupied by the soil. Five repetitions were carried out In order to obtain a precise result. The actual density is given by the formula

$$dr = \frac{Ps}{V}$$

With :

- dr: Real density (g/cm3).
- Ps: dry weight (g).
- Vs: volume of water displaced (cm)^3

Porosity is calculated from the results obtained by apparent density as follows:

$$n = \frac{da}{dr}$$

With :

- n%: total porosity.
- da: apparent density (g/cm3).
- dr: true density (g/cm3).

2.3.4. Determination of infiltration speed

To determine the rate of water infiltration into the ground, we used a method based on a double-ring infiltrometer fig (29). The rings used had specific dimensions: an outer ring with a diameter of 50 cm and an inner ring with a radius of 30 cm, welded together. We first sank the rings into the ground so that they were level and in close contact with the ground, to avoid any water leaks. Then we filled both rings simultaneously with water until the initial level in both rings was equal, thus saturating the soil. Once the water level had been reached, we started the stopwatch and noted the initial water level using a graduated ruler. At regular intervals, every 5 mm, we measured and recorded the drop in water level in the the outer ring, indicating that the soil had reached a stable infiltration rate. The infiltration rate was calculated by dividing the height of water infiltrated by the elapsed time

Figure 29: Double-ring infiltrometer

The infiltration rate of water is obtained by the following relationship :

$$\boldsymbol{Vitesse\ d'infiltration} = \frac{\boldsymbol{Profondeur\ (cm)}}{\boldsymbol{Temps\ (min)}}$$

2.3.5. Determination of microbial activity

For microbiological analysis, soil samples are taken from the surface layer (0-20 cm), which is the most biologically active. In order to obtain a representative sample, six samples of 500 g each are taken with an auger, spread diagonally over an area of 100 m², after removing surface debris. The samples are weighed in sterile sampling containers fig 30. The samples are then transported to the laboratory under controlled conditions to guarantee their microbiological integrity. This formulation is more precise and focuses on the importance of transport under appropriate conditions, which is crucial to the validity of subsequent microbiological analysis, preserving their microbiological integrity.

Figure 30: Sampling boxes

- **Preparation of biological material**

Any soil sample prepared for a micro-organism study is first crushed in a mortar. It is then passed through two successive sieves: the first with 5 mm holes, then the second with 2 mm holes. This process removes elements that are too large and unwanted organic debris.

- **Enrichment and enumeration of total aerobic mesophilic soil flora**

A 2 g sample of soil is aseptically added to 40 ml of sterile Luria-Bertani (LB) medium in a 250 ml bottle. This suspension, constituting the stock solution (10^{-1} dilution), is then incubated at 30°C for 24 hours with constant stirring at 150 rpm (Guo et al., 2010).

Figure 31: Stages in the enrichment of total aerobic mesophilic soil flora

- **Performing decimal dilutions**

Analysis of the total aerobic mesophilic flora (FAMT) was carried out using the suspension-dilution method as described by (Rayilli, 1968). After incubation of the enriched culture, a series of decimal dilutions (from 10^{-1} to 10^{-9}) was prepared as follows:

1. 1 ml of the stock suspension (10^{-1}) is aseptically transferred to 9 ml of sterile physiological water, giving the 10^{-2} dilution.
2. After homogenisation for two minutes, 1 ml of this dilution is taken and added to 9 ml of sterile physiological water, producing dilution 10^{-3}.
3. This process is repeated until the 10 % dilution is obtained.$^{-9}$

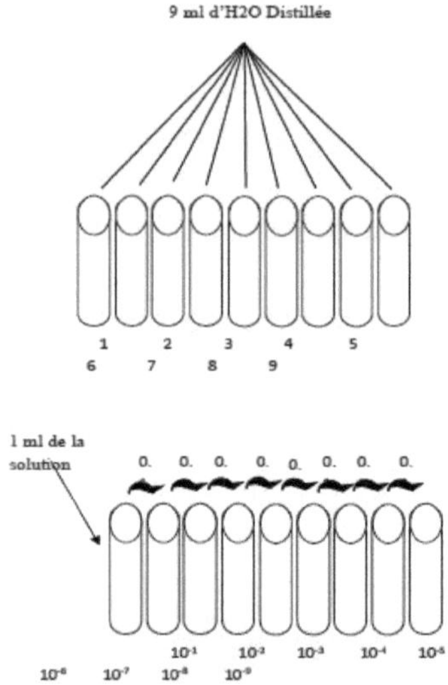

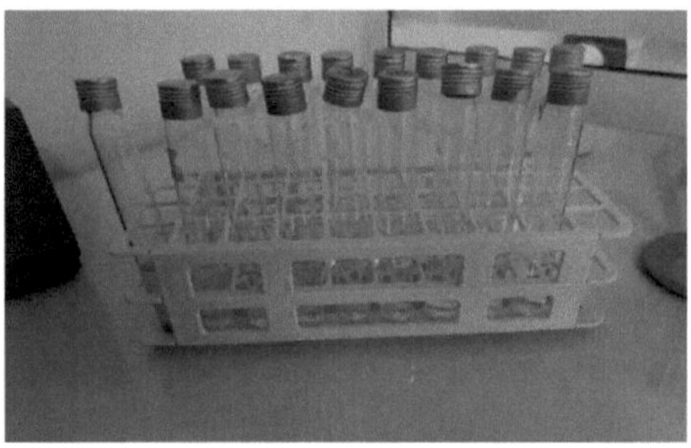

- **Seeding**

The technique used in this study to isolate edaphic microorganisms is based on the surface inoculation method. Serial dilutions of the soil sample are prepared and then inoculated onto a solid culture medium, in this case nutrient agar (NA). The experimental protocol is as follows: a volume of 0.1 ml is taken from each dilution (from 10^{-3} to 10^{-9}) using a sterile micropipette. This inoculum is then placed in the centre of a Petri dish containing the previously solidified GN medium. The inoculum is spread evenly over the surface of the medium using a sterile rake. Subsequently, the seeded Petri dishes are inverted and incubated at a temperature of 30°C. The incubation period extends from 24 to 72 hours (Guo et al., 2010), allowing microbial colonies to grow and develop.

Figure 33: Isolation of microorganisms

- **Enumeration of total mesophilic aerobic flora**

The total aerobic mesophilic edaphic flora is counted in accordance with the protocols established by Marchal and Bourdon (1982). Determination of microbial density, expressed in colony-forming units per gram of soil (CFU/g), is calculated as the arithmetic mean of the counts, taking into account the dilution factors applied. This methodology can be formalised by the following equation, designated as formula 01 :

Formula 01: N (CFU/g) = $(\sum n) / (\sum d \times V)$ Where:
N: Number of microorganisms in CFU/g.

n: Number of colonies counted.

V: Volume deposited (1 ml).

d: Dilution

2.4. Methodology for crop-related measurements

The methodology for analysing the effects of the two techniques on crop development focused respectively on :

- Lift rate
- The tillering rate
- The height of the plant.

- Root development
- Root density.
- Performance.

2.4.1. Determination of emergence rate and tillering rate

To determine the emergence rate and tillering rate, we used a wooden frame measuring ¼ square metre, and counted the number of plants to arrive at the number of germinated seeds per hectare.

Figure 4: Determination of barley tillering rate in the field

2.4.2. Determining stem heights and diameters

To measure the height of the plants, we used a measuring tape and followed the evolution of the aerial part of the plant to arrive at the height corresponding to the cut.

Figure 5: Determination of stem heights and diameters

2.4.3. Determining root density

To measure root elongation in barley, we used the coring method. A metal cube with sharpened ends is used to measure root elongation. used to remove the plants with their roots. This sample (soil + roots) is placed in a fine mesh sieve and the soil is removed using a jet of water. The samples are then soaked in water for 24 hours and washed again until all the soil has been removed. They are then dried and weighed.

Figure 6: Root removal

2.4.3. Determining the estimated yield

To estimate the grain yield on the different plots, we took two measurable characteristics, namely the number of grains per square metre and the weight per thousand grains. The relationship used to calculate the yield is as follows:

$$\text{Rendement estimé} \left(\frac{Qx}{ha}\right) = \text{nombre de grains par } m^2 \times \frac{PMG}{10000}$$

3.5. Results and discussion

2.5.1. Combined effect of soil preparation techniques and compaction on the physico-mechanical behaviour of the soil

2.5.1.1. Effect on humidity

The first soil parameter studied is moisture, which is of vital importance for the germination, emergence and development of wheat plants. Soil moisture is determined by the amount of rainfall accumulated during the trial period. The moisture levels measured on the plot throughout the campaign are derived from the rainfall recorded during monitoring, and reflect the water retention capacity of the topsoil, i.e. the effectiveness of the structure in retaining water.

Stage 1: (samples taken after precipitation)

The results of the variation in soil water content are shown in the following figures:

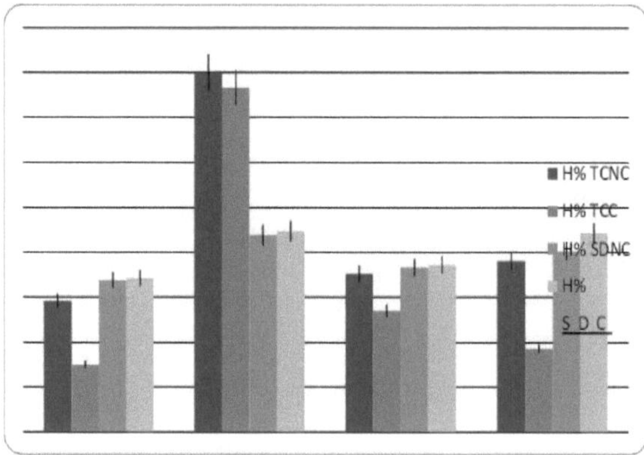

Figure 7: variation in soil moisture in relation to depth and compaction at the emergence stage

From this figure, we can see that in the first stage, in all horizons, we recorded a difference in water content between the non-compacted areas and the areas compacted using the conventional tillage technique. The water content decreases in the areas compacted using conventional tillage, and this decrease is noticeable in all horizons, being more significant in horizons 1 and 4, with a difference of 8% and 7% respectively. It can also be seen that in the second horizon, we recorded the highest water content values for conventional working, in the compacted and uncompacted zones.

Stage 2

The results of the variation in soil water content are shown in the following figure:

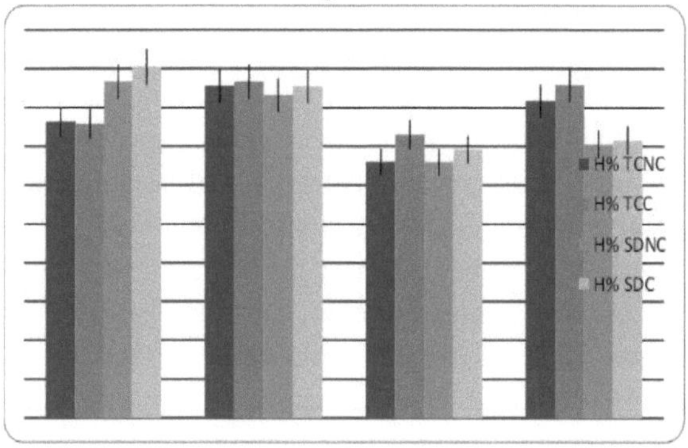

Figure 8: Variation in soil moisture in relation to depth and compaction at tillering stage

At this stage, we have recorded a period of drought, and the results show that there is no great difference in the water content of the soil. water content between compacted and non-compacted areas for both cultivation techniques, and in all horizons.

Stage 3: (sampling after 15 days of rainfall in April)

The results of the variation in soil water content are shown in the following figure:

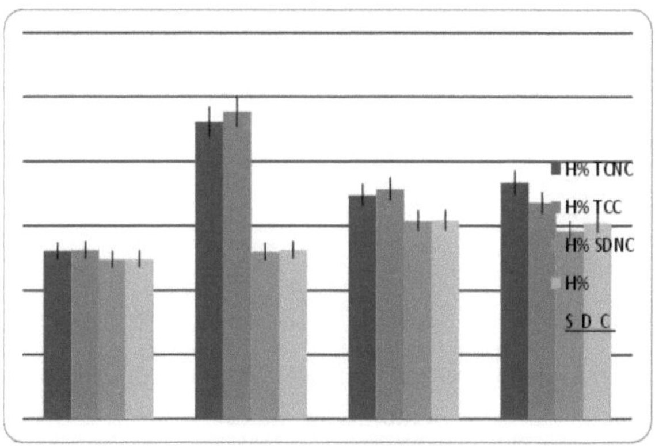

Figure 9: Variation in soil moisture in relation to depth and compaction at the heading stage

At this stage, we took samples after 15 days of rainfall in April. The results showed that compacted soils dry out quickly, which can be explained by the high temperatures recorded during this period. We found that the second horizon of the conventional plots had the highest moisture content. This result can be explained by the presence of a compact layer on the surface, which dries out quickly, becoming an insulating layer and preventing the evaporation of the water stored in the second horizon.Generally speaking, compaction in conventionally worked soils has an effect on soil water content. The results show that ploughing allows an increase in water content in the first horizon, but in compacted areas, this is not the case; the soil is compacted and dries out quickly. We also note that soil moisture decreases slightly in the second horizon of compacted areas, thanks to the compacted surface layer that prevents evaporation. We also conclude that conventional tillage is more susceptible to compaction than direct seeding, because the soil becomes lighter and wetter after turning in conventional tillage, which facilitates compaction after passes by farm machinery, unlike direct seeding, where the soil retains its structure and resists passes by machinery.

3.5.1.2. Effect on porosity

The second parameter analysed is porosity, which largely determines water retention by the soil, aeration of the roots and their progress through the soil. Analysis of the combined effect of cultivation techniques and soil compaction on bulk density and porosity is shown in the following figures:

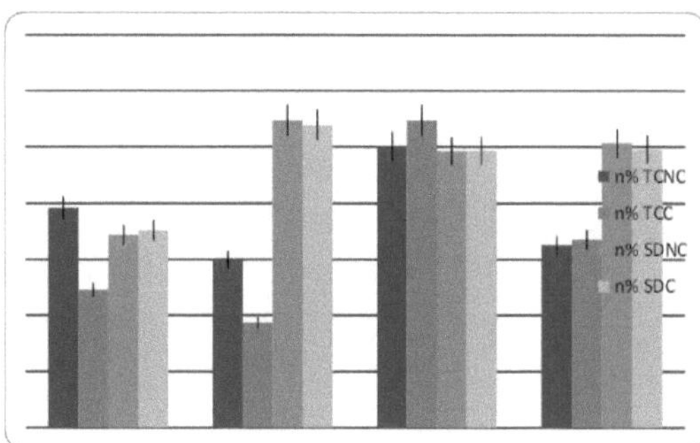

Figure 10: Variation in soil porosity in relation to depth and compaction at the emergence stage

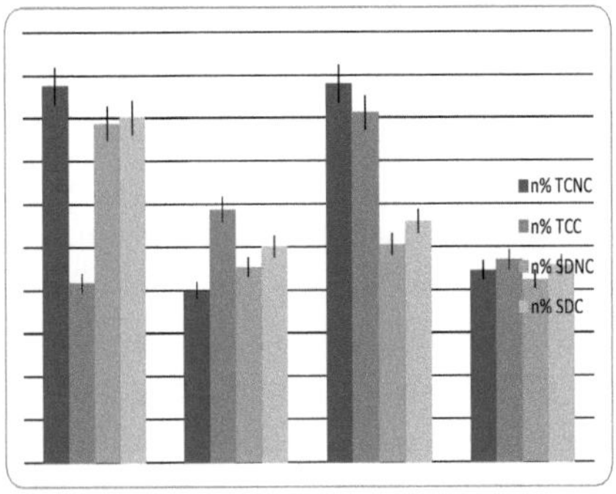

Figure 11: Variation in soil porosity in relation to depth and compaction at the tillering stage

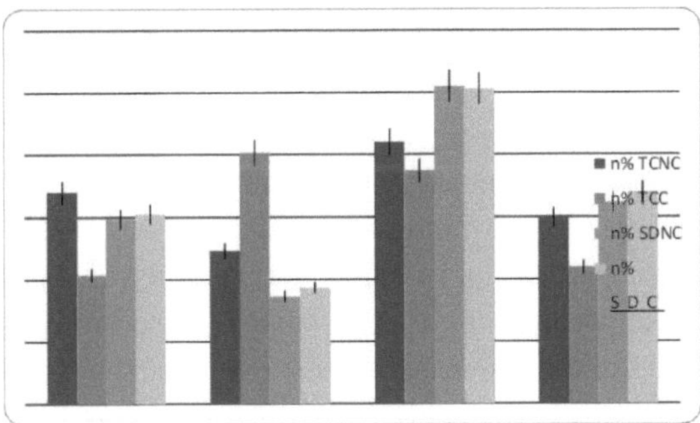

Figure 12: Variation in soil porosity in relation to depth and compaction at the heading stage

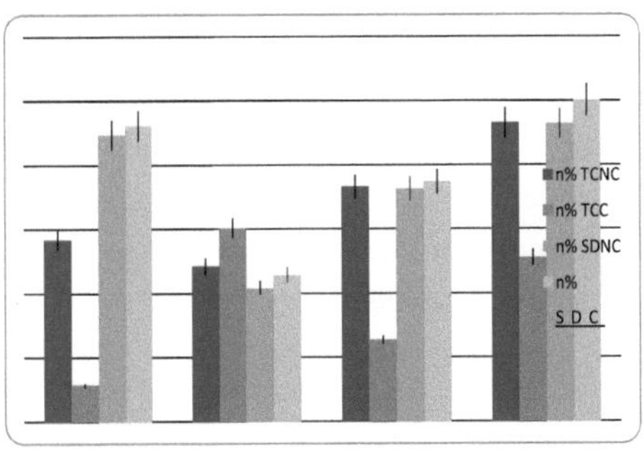

Figure 13: Variation in soil porosity in relation to depth and compaction at the ripening stage.

Porosity is a physical characteristic that reflects the structural state of the soil; a soil whose aggregates are distributed in such a way that it resists the stresses associated with the wetting/drying or freezing/thawing cycle is said to have a stable structure, and therefore maintains its pore space at a fixed level. In addition, a particulate structureis the opposite of a fragmentary structure, highly unstable and remains sensitive to multiple degradations such as capping or compaction caused by the repeated passage of heavy machinery. It is the stability of the structure that undeniably affects the state of the soil, including the pores.In our case, and after ploughing, we recorded an increased increase in porosity, with a deviation of 10% from the initial value. After compaction by several passes of the tractor in ploughed areas, porosity decreased again, with a deviation of 15% in the first horizon (surface layer) and 10% in the second horizon. This can be explained by the type of porosity existing in the ploughed soil, which is the porosity This type of porosity is very sensitive to the loads exerted by agricultural machinery and the passage of machinery over the land. In ploughed plots, for horizons between 20 and 30 cm, we recorded lower porosity values between compacted and non-compacted areas.However, with direct seeding, there is a slight decrease in porosity between compacted areas (compacted by several passes of the tractor) and non-compacted areas for direct seeding in all horizons. This can be explained by the type of porosity present in the soil. With direct seeding, there is no tillage, the porosity is tubular and is linked to the biology of the soil (earthworms, root penetration, etc.). This type of porosity is less sensitive to the loads exerted by agricultural machinery and the passage of machinery over the land. Overall, we can conclude that, for conventional tillage, the degree of porosity increases sharply after ploughing, and decreases sharply after compaction, especially in the surface layers. Porosity also increases slightly at various stages in the deeper horizons as the roots develop.On the other hand, in direct seeding, we recorded almost identical values in the compacted and non-compacted zones, with the porosity value increasing slightly with the development of the root system.We can conclude that the passage of farm machinery in conventional tillage contributes to soil compaction in the superficial layers, and has little effect on no-till zones.

2.5.1.3. Effect on soil penetrometric resistance

Measuring penetrometric resistance using a penetrometer is one of the indicators frequently used to give an idea of the state of soil compaction and the constraints the soil places on root growth. Observations on the evolution of the penetrometric resistance of the soil in relation to the number of passes made by agricultural machinery and the depth are presented in the following figures:

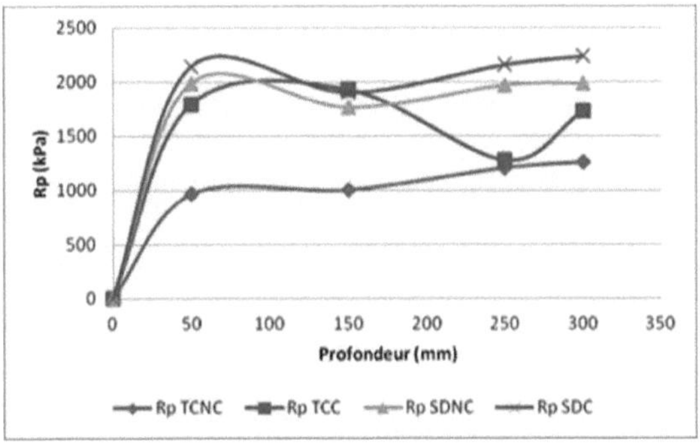

Figure 14: variation in soil penetrometry in relation to depth and compaction at the emergence stage

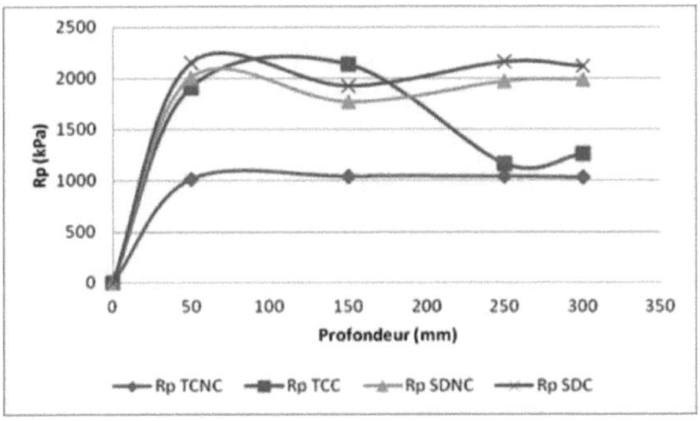

Figure 15: variation in soil penetrometry in relation to depth and compaction at the tillering stage

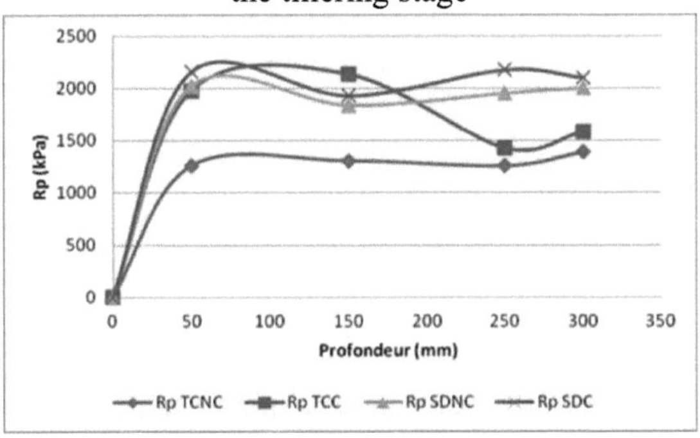

Figure 49: Variation in soil penetrometry in relation to depth and compaction at the heading stage

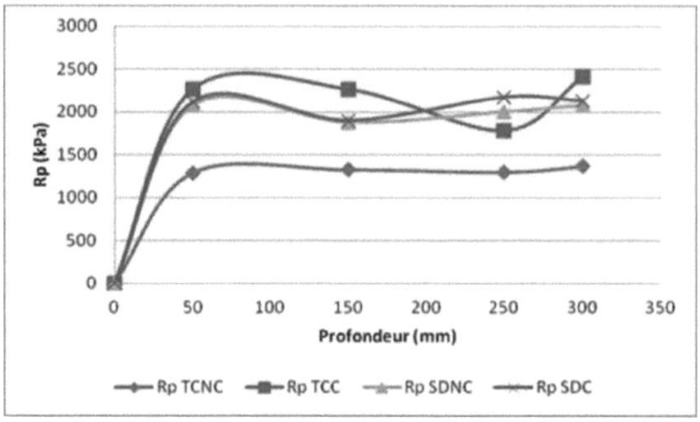

Figure 16: Variation in soil penetrometry in relation to depth and compaction at the ripening stage

The results obtained show that the penetrometric resistance of the soil is modified by the increase in the number of passes made by the agricultural machinery used and the cultivation technique used. We found that after each operation or pass :

➢ Areas **not compacted** by **conventional tillage**

Penetrometric resistance falls sharply after ploughing, and rises slightly after each pass of the tractor during tillage (around 1000 kpa), and again after the passage of the sprayer, and also after the passage of the combine harvester and pick-up baler (around 1400 kpa). Penetrometric resistance increases slightly after each pass of the tractor and agricultural machinery, especially in the shallow horizons (between 0 and 10 cm). For the deeper horizons, between 20-30 cm, the penetrometric resistance values are higher, but with little change after each passage of machinery.

➢ Areas **compacted** using **conventional methods**

The value of penetrometric resistance is greater after compaction, reaching high values in the first horizon after each passage of a tractor or agricultural machine. The initial value before compaction does not exceed 1000 kpa, then it increases progressively after compaction (1700 kpa), after the passage of the combine harvester at harvest time, and the passage of the pick-up baler, it reaches a maximum value of around 2500 kpa. For the deeper horizons, between 20-30 cm, the penetrometric resistance values in the compacted areas are higher than those in the non-compacted areas, but with little change after the passage of farm machinery over the land compared with the surface areas. In addition, in direct seeding, we recorded almost identical penetrometric resistance values in compacted and non-compacted areas, with a slight increase in resistance after each pass of the machinery in the first horizon only. For horizons in At a depth of 20-30 cm, the penetrometric resistance values are higher, with very little change after the machines have been used. In conclusion, compaction mainly affected the assembly porosity resulting mainly from tillage. This is in contrast to direct seeding, where the soil is more resistant to compaction thanks to its natural vertical porosity (tubular porosity). These results were confirmed by (Vincent Tomis 2021) in an oral communication, as part of an Agrotransfer Hauts de France agricultural research and development project.

Statistical analysis

3.6.1. Statistical analysis of changes in moisture, porosity and penetrometric resistance as a function of cultivation technique and soil compaction

Statistical analysis of changes in moisture, porosity and penetrometric resistance as a function of cultivation technique and soil compaction yielded the following results:

Table 9: Results of statistical analyses :

	H (%)		n (%) (cm).		Rp (kPa)	
	p-value	Group s	p-value	Group s	p-value	Group s
TCNC	NS	A	NS	A	0,008 HS	B
CBT		A		A		A
SDNC		A		A		A
SDC		A		A		A

NS: Not significant/HS: Highly significant

3.6.2. The relationship between Rp, H% and n%.

The multiple regression analysis between the dependent variable Rp and the independent variables H%, n% yielded the following results:

Table 10: Analysis of variance for Rp, H% and n%.

Source	DDL	Sum of squares	Average square	F	Pr > F
Model	2	41810,271	20905,136	0,954	0,0410
Error	13	284789,338	21906,872		
Corrected total	15	326599,609			
Calculated against the model Y=Average(Y)					

Firstly, the analysis of variance table, which is a result that needs to be analysed carefully. This is where we test whether we can consider that the selected explanatory variables (n) and (H%) contribute a significant amount of information to the model or not. In other words, it is a way of testing whether the mean of the variable to be modelled Rp would be sufficient to describe the results obtained or not. The Fisher F test is used. Given that the probability associated with the F in this case is less than 0.04, this means that we take a risk of being wrong of less than 4% in concluding that the explanatory variables contribute a significant amount of information to the model. Secondly, we need to check that there is no significant correlation

2.5.1.4. Effect on infiltration speed

The speed at which water infiltrates the soil is of vital importance in analysing the characteristics of our soil, and determining the soil's useful reserves so that irrigation water doses can be adapted.

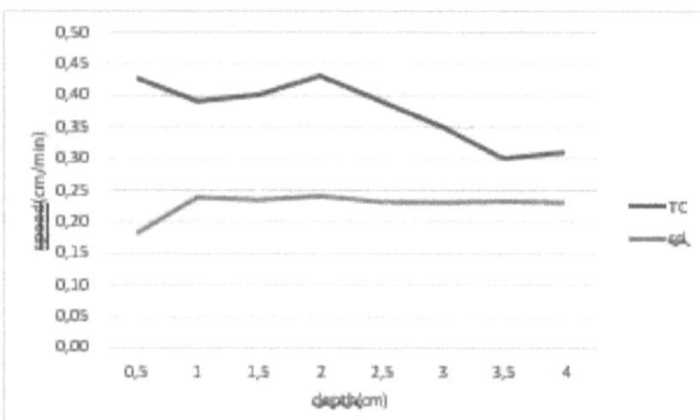

Figure 51: Infiltration speed (cm/min)

The graph above shows water infiltration rate as a function of soil depth for the two soil management systems, direct seeding (DS) and conventional tillage (CT).It is clear that the infiltration rate into the soil by the CT is initially higher (around 0.42 cm/min at a depth of 0.5 cm) but gradually decreases as the depth increases. The SD has a lower and relatively constant infiltration rate of around 0.20 to 0.22 cm/min over the entire depth measured. Based on these results, we can conclude that a ploughed and turned soil offers good infiltration of irrigation water compared with unworked soil (direct seeding), even if it is more sensitive to compaction caused by the passage of machinery on the surface. The impact of compaction can be seen in the deep horizons, where we recorded lower infiltration speeds than at the surface.

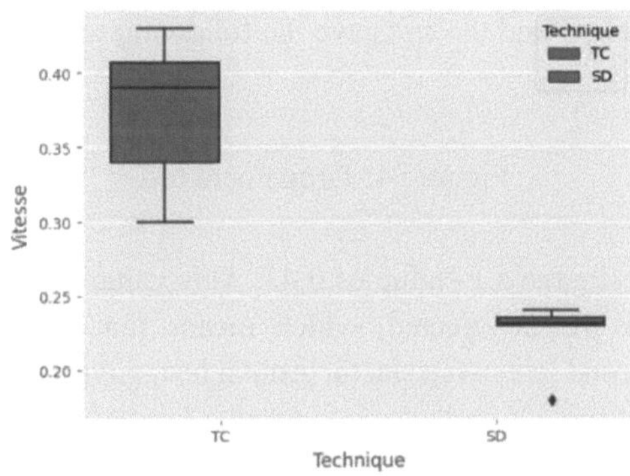

Figure 52: Comparison of median water infiltration speed between conventional and no-till systems

- **Statistical analysis**

➢ **Normality test**

A normality test was carried out to verify the normal distribution of our data on variations in water velocity in the experimental trial as a function of tillage techniques. The graphs below show the distribution of soil water infiltration velocity data.

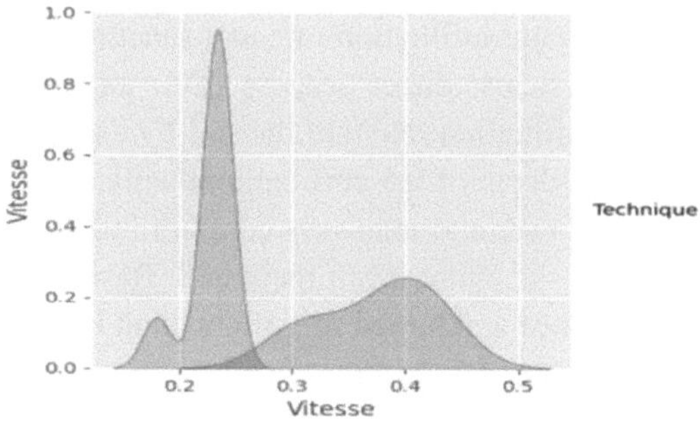

Figure 53: Figure: Variation in infiltration rate as a function of tillage technique

For the normality test, we applied the Jarque_bera test, using the Python programming language, and the test gave the following results following :

```
In [6]: stats.jarque_bera(df['Vitesse'])

Out[6]: SignificanceResult(statistic=1.6523022980202953, pvalue=0.4377308089380214)
```

Figure 54: Jarque bera test

The Jarque_bera test gave a P-value of 0.43. This value is greater than 0.05, so the null hypothesis H_0 is rejected, which means that the data are normally distributed. The second test, a one-factor Anova test, gave the following results:

```
In [7]: mod=ols('Vitesse ~ Technique', data=df).fit()
        aov_table=sm.stats.anova_lm(mod,typ=2)
        print(aov_table)

             sum_sq    df          F    PR(>F)
Technique  0.087271   1.0  61.551555  0.000002
Residual   0.019850  14.0        NaN       NaN
```

Figure 55: Anova test

The analysis of variance is a result that must be analysed carefully. It is at this level that we test whether we can consider that the explanatory variable selected (infiltration rate) provides a significant amount of information to the model or not. In other words, it is a way of testing whether the average of the variable to be modelled (infiltration rate) is sufficient to describe the results obtained or not. Given that the p-value is less than 0.05, a highly significant result (0.000002), we can say that the tillage technique has a very strong impact on the speed of water infiltration into the soil.The results of our trial on the speed of water infiltration and the statistical analyses confirm that conventional tillage favours water infiltration compared with direct seeding, even though CT is more susceptible to compaction because of the high number of times machinery passes over the plot (ploughing, ploughing over, shallow cultivation, sowing, phytosanitary treatment, etc.). This can be explained by the good porosity of ploughed soils, whereas no-till soils are compact and prevent water infiltration.

2.5.1.5. Effect on microbial activity

The histogram below compares the number of bacteria in the soil, expressed in colony-forming units per millilitre (cfu/ml), under two management conditions: conventional tillage (CT) and semi-direct tillage (SD). The vertical axis represents the number of bacteria (cfu/ml), ranging from 0 to 10.

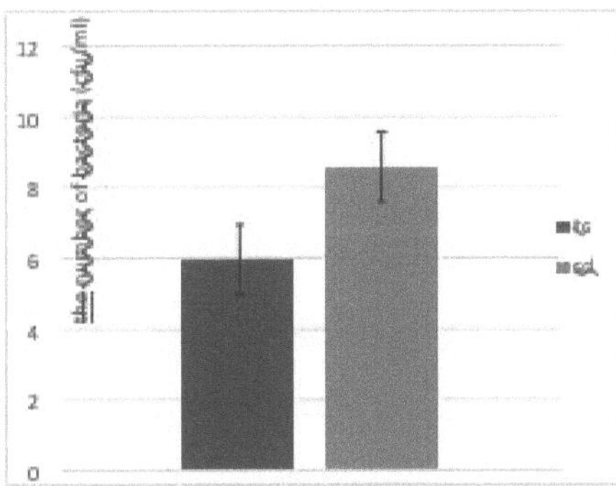

Figure 56: Bacteria count

He groundin SD condition contains around 9 cfu/ml of bacteria, while the soil in TC condition contains around 6 cfu/ml. This significant difference of 3 cfu/ml. It's clear that no-till favours microbial activity over conventional tillage.

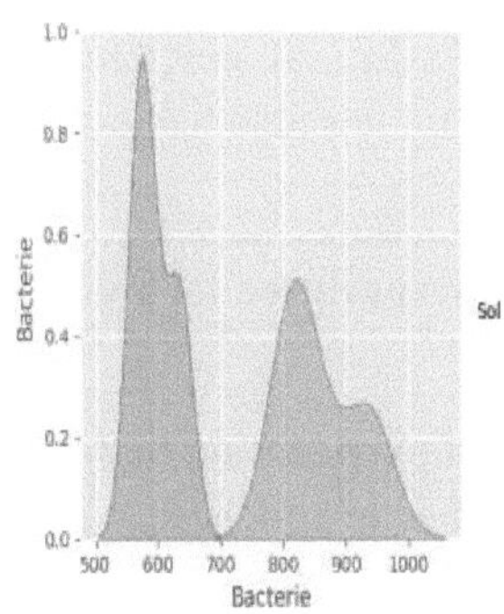

Figure 57: Distribution of bacterial count data for the experimental trial

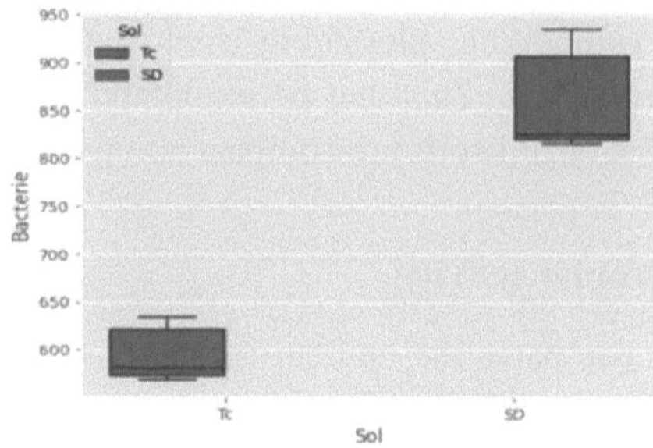

Figure 58: Comparison of the median number of bacteria between conventional and no-till systems

- **Statistical analysis Anova test :**

```
mod=ols('Bacterie ~ Sol', data=df).fit()
aov_table=sm.stats.anova_lm(mod,typ=2)
print(aov_table)

                 sum_sq    df          F    PR(>F)
Sol       208560.333333   1.0  94.233324  0.000002
Residual   22132.333333  10.0        NaN       NaN
```

Figure 59: Anova test

The analysis of variance table is a means of testing whether the mean of the variable to be modelled, the number of bacteria in the soil, is sufficient to describe the results obtained or not. Given that p-value is less than 0.05, we can say that the tillage technique has a very strong impact on soil bacteria counts.

From the results obtained, we can conclude that the cultivation technique with or without ploughing affects the soil's structure, water content, nutrient availability and temperature, all of which influence bacterial activity. Ploughing disrupts soil structure by turning over the top layers, which disturbs microorganisms and leads to a long-term reduction in microbial diversity. Direct seeding, on the other hand, preserves the soil's natural structure, maintaining distinct layers of organic matter and promoting more stable habitats for bacteria.

2.8.Combined effect of soil preparation techniques and soil compaction on plant development

2.8.1. Combined effect of soil preparation techniques and soil compaction on emergence rate and number of spikes/sq. m.

The results of the combined effect of soil preparation techniques and soil compaction on Tallage/m^2 and ear/m^2 are presented in the following figures:

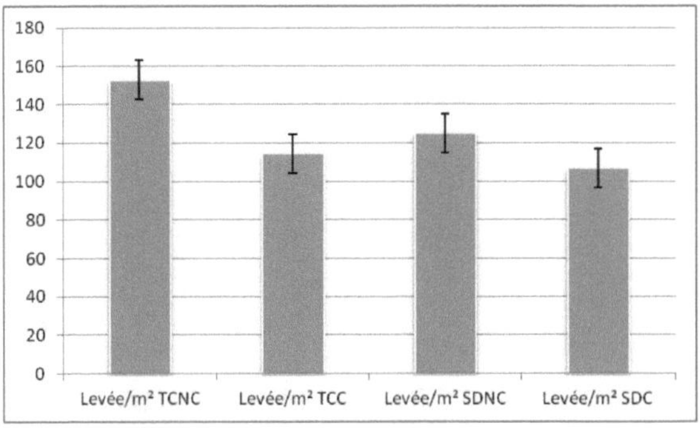

Figure 60: Combined effect of soil preparation techniques and soil compaction on emergence rate

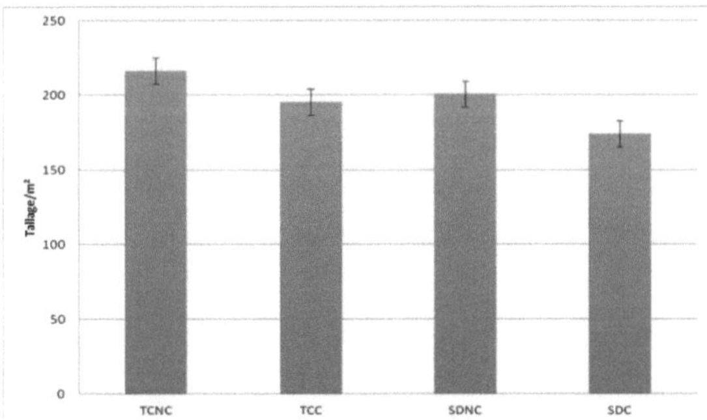

Figure 61: Combined effect of soil preparation techniques and soil compaction on tillering/m².

The results clearly show that cultivation technique and compaction have an effect on emergence rate and the number of ears per square metre. Conventionally tilled, non-compacted soils gave good tillering compared with compacted, tilled soils, where there was a difference of 30 plants/m² between the two cases. The number of spikes in non-compacted and compacted conventionally tilled soils was around 220 plants/m² and 190 plants/m² respectively. For direct seeding, the observation is the same, since the tillering rate and the number of plants/m² is clearly higher in non-compacted soils, where we recorded 200 plants/m², and 165 plants/m² in compacted areas.

2.8.2. Combined effect of soil preparation techniques and soil compaction on stem length and diameter

Combined effect of soil preparation techniques with soil compaction along the length shown in the following figure :

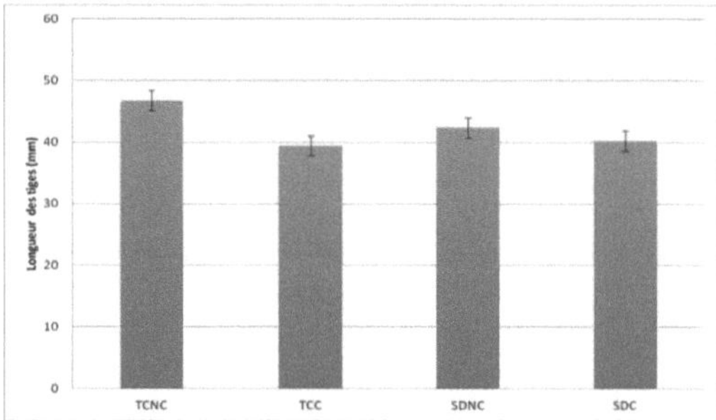

Figure 62: Combined effect of soil preparation techniques and soil compaction on stem length

The combined effect of soil preparation techniques and soil compaction on the diameter of the ears is shown in the following figure:

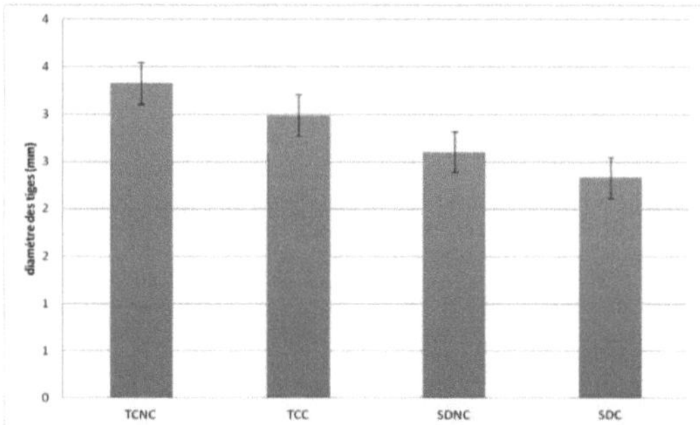

Figure 63: Combined effect of soil preparation techniques and soil compaction on ear diameter

The results obtained in the field show that there is an effect on stem length and diameter. The latter are clearly greater in conventionally worked, non-compacted soils, with values in non-compacted and conventionally worked compacted soils of around 47 cm, 39 cm and 3.3 mm, 2.95 mm respectively. For direct seeding, the observation is the same, since the length of the stems is clearly greater in non-compacted soils than in compacted soils, where we recorded 42.5 cm, 40 cm and 2.6 mm 2.2 mm respectively.

2.8.3. Combined effect of soil preparation techniques and soil compaction on the number of grains per ear and the weight of 1000 grains

The combined effect of soil preparation techniques and soil compaction on the number of grains per ear is shown in the figure below:

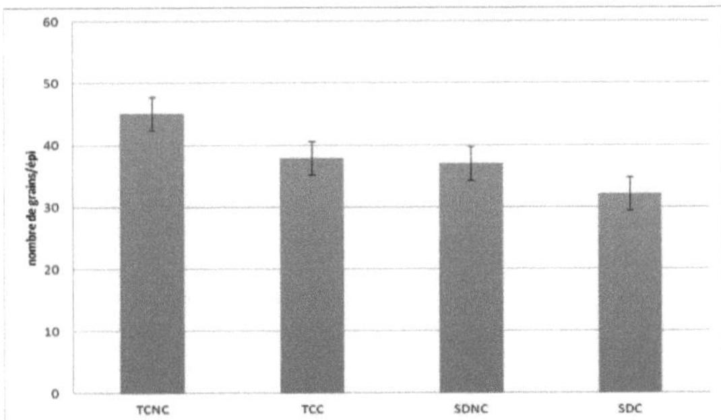

Figure 64: Combined effect of soil preparation techniques and soil compaction on the number of grains per ear

The combined effect of soil preparation techniques and soil compaction on the weight of 1000 grains per ear is shown in the following figure:

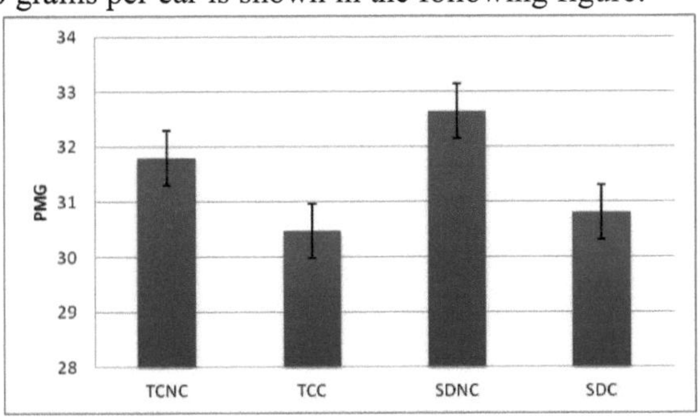

Figure 65: Combined effect of soil preparation techniques and soil compaction on the weight of 1000 grains.

The first thing we can see in the two figures above is the effect of cultivation technique and compaction on the number of grains per ear and the thousand grain weight. For the first figure, it appears that the number of grains per ear is slightly better for the plots worked and not compacted, the average value recorded for the TCNC and TCC is 45 grains/ear and 37 grains/ear respectively, while that obtained for direct seeding is 36 grains/ear and 32 grains/ear in the plots not compacted and compacted respectively. In the second figure, it can be seen that the weight of 1000 grains is much better for plots that have been worked and those that have not been compacted. The average value recorded for TCNC and TCC is 32.6 g and 30.8 g respectively, while that obtained for direct seeding is 33.1 g and 31.5 g in plots that have not been compacted and those that have been compacted respectively.

CONCLUSION

In conclusion, we can say that the tillage technique and the compaction of agricultural soils have a direct effect on the development of barley and its root growth.In non-compacted tilled plots (TCNC), soil structure is mainly created by tillage operations, whereas in no-till systems (direct seeding) structure is mainly created by the action of the climate (on the surface in temperate regions) and by biological processes.However, the compaction artificially created in ploughed and no-till plots manifests itself in superficial layers of soil that dry out quickly and have a greater resistance to penetration, which prevents good emergence and tillering, and also hinders good plant development on the surface, as well as preventing good root development compared with ploughed, non-compacted soils. This can lead to a series of unfavourable growing conditions during the crop development period in compacted soils. One of the essential roles of tillage, therefore, is to facilitate root growth and plant development.We can conclude that conventional tillage disturbs the soil structure by breaking up the aggregates, which encourages rapid infiltration of water to the surface. This corresponds to the initial increase in speed observed. However, for deeper horizons, compaction under the worked zone can occur, reducing soil porosity and further limiting water penetration. This explains the gradual decrease in water infiltration speed. Direct seeding favours a greater abundance of bacteria compared with conventional tillage. Although the increase in the bacterial population is generally considered beneficial, its impact on agricultural productivity can vary. Studies have shown that the effects of semi-direct seeding on yields depend on climatic conditions and management practices. However, a structure with low porosity due to compaction after ploughing has a negative effect on plant and crop root development and yields.

REFERENCES

Aikins, S., Afuakwa, J., 2012. Effect of four different tillage practices on soil physical properties under cowpea. Agric. Biol. J. N. Am. 3, 17-24.

Amara, M., Boudhar, L., Adli N., Feddal, M.A., 2008. Evolution of soil penetrometric resistance in relation to moisture and porosity, under the action of the working parts of a conventional soil preparation line.
Vol. 2, n° 1 et 2 - January-December 2008, Science et technique, Sciences appliquées et Technologies. Burkina Faso.

Augustin, K., Michael, K., Joachim, B., Rainer, D., 2020. Wheel Load and Wheel Pass Frequency as Indicators for Soil Compaction Risk: A Four-Year Analysis of Traffic Intensity at Field Scale. Geosciences 2020, 10(8), 292; https://doi.org/10.3390/geosciences10080292.

Batey, T., 2009. Soil compaction and soil management - a review. Soil Use and Management, December 2009, 25, 335-345.

Bessou, C., Ferchand, F., Gabrielle, B., Mary, B., 2010. Biofuels, green house and climate change. A review. Agronomy for sustainable development. 31. 1-79.

Bessou, C., Mary B., Leonard, J., Roussel, M., Gréhan, E., Gabrielle, B., 2010. Modelling soil compaction impacts on nitrous oxide emissions in arable fields. European Journal of Soil Science 61 (3), 348-363.

Botta, G.F.; Tolón-Becerra, A.; Bienvenido, F.; Rivero, D.; Laureda, D.A.; Ezquerra-Canalejo, A. 2018; Contessotto, E.E. Sunflower (Helianthus annuus L.) harvest: Tractor and grain chaser traffic effects on soil compaction and crop yields. Land Degrad. Dev. 29, 4252-4261.

Chabane, M., Boussard, J., 2012. Cereal production in Algeria: From today's realities to tomorrow's strategic perspectives. [Other], 20 p.

Derpsch, R., Friedrich, T. 2009. Development and Current Status of No-till Adoption in the World. In Proceedings of the 18th Triennial Conference of the International Soil Tillage Research Organization (ISTRO), Proceedings on CD, Izmir, Turkey, 15-19 June; pp. 1-13.

Ernst, G., Emmerling, C., 2013. Impact of five different tillage systems on soil organic carbon content and the density, biomass, and community composition of earthworms after a ten year period. Eur. J. Soil Biol. 45, 247-251.

Guy, R., 2012. Evaluation and control of compaction risk, from plot to national territory. Innovations Agronomiques, INRAE, 21, pp.45-57.

Lefebvre, A., 2010. Spatialization of hydromechanical soil functioning models applied to the prediction of settlement risks at the scale of France. Thése Doc. INRA Orléns. 267 P.

Moinfar, A., Shahgholi, G., Abbaspour-Gilandeh, Y., Herrera-Miranda, I., Hernández-Hernández, J.L., Herrera-Miranda, M., 2021. Investigating the Effect of the Tractor Drive System Type on Soil Behavior under Tractor Tires. Agronomy, 11, 696.

Nawaz M., Bourrié G., Trolard F., 2013. Soil compaction impact and modelling. A review. Agronomy for Sustainable Development;33:291-309.

Orzech, K., Załuski, D., 2020. Effect of companion crops and crop rotation systems on some chemical properties of soil. J. Elem, 25, 931-949.

Roger-Estrade J., Adamiade V., Arrouays D., Bartoli M., Baranger E., et al., 2011. Physical degradation of agricultural and forest soils due to compaction : main results of the GESSOL ADD-DST project. Etude et Gestion des Sols, Association Française pour l'Etude des Sols, 18 (3), pp.187-199.

Schjønning, P., (2018). The role of soil organic matter for maintaining crop yields: evidence for a renewed conceptual basis. In: Advances in agronomy, vol. 150, Elsevier, New York, pp 35-79.

Shah, A., Tanveer M., Shahzad B., Yang G., Fahad S., Ali S., Bukhari A., Tung A., Hafeez, A., Souliyanonh, B., 2017. Soil compaction effects on soil health and crop productivity: an overview. Environmental Science and Pollution Research;24:10056-10067

Somasundaram, J.; Chaudhary, R.S.; Awanish Kumar, D.; Biswas, A.K.; Sinha, N.K.; Mohanty, M.; Hati, K.M.; Jha, P.; Sankar, M.; Patra, A.K.; et al, 2018. Effect of contrasting tillage and cropping systems on soil aggregation, carbon pools and aggregate-associated carbon in rainfed Vertisols. Eur. J. Soil Sci. 69, 879-891.

Wozniak, A., 2020. Effect of various tillage systems on winter barley yield, weed infestation and soil properties. Appl. Ecol. Environ. Res. 18, 3483-3496.

MIX
Papier aus verantwortungsvollen Quellen
Paper from responsible sources
FSC® C105338

Printed by Books on Demand GmbH, Norderstedt / Germany